THE IRON
INDUSTRY

Richard Hayman

SHIRE PUBLICATIONS
Bloomsbury Publishing Plc

Kemp House, Chawley Park, Oxford OX2 9PH, UK
29 Earlsfort Terrace, Dublin 2, Ireland
1385 Broadway, 5th Floor, New York, NY 10018, USA
Email: shire@bloomsbury.com
www.shirebooks.co.uk

SHIRE is a trademark of Osprey Publishing Ltd

First published in Great Britain in 2016
Transferred to digital print on demand in 2023

A CIP catalogue record for this book is available from the
British Library.

Shire Library no. 817
Print ISBN: 978 0 74781 483 2
ePDF: 978 1 78442 084 0
ePub: 978 1 78442 083 3

Typeset in Garamond Pro and Gill Sans
Printed and bound in Great Britain

MIX
Paper | Supporting
responsible forestry
FSC   FSC® C013604
www.fsc.org

The Woodland Trust
Shire Publications supports the Woodland Trust, the
UK's leading woodland conservation charity.

www.shirebooks.co.uk
To find out more about our authors and books visit
our website. Here you will find extracts, author
interviews, details of forthcoming events and the
option to sign-up for our newsletter.

COVER IMAGE
Front cover design and photography by Peter Ashley,
showing details of the Iron Bridge, opened in 1781
over the River Severn in Shropshire. Back cover: Detail
from a lathe at the iron works at Blists Hill, Shropshire,
photography by Peter Ashley.

TITLE PAGE IMAGE
Nails were a staple product of the medieval iron industry,
made by shaping and cutting thin rods under a hammer.
This illustration was published in Germany in 1568.

CONTENTS PAGE IMAGE
Cyfarthfa Ironworks in 1811, when it was the largest in
the world. On the left are the blast furnaces and casting
houses and on the right are the puddling furnaces, behind
which are workmen's houses.

ACKNOWLEDGEMENTS
Permission to reproduce illustrations has been given
by the following: Cadbury Research Library, Special
Collections, University of Birmingham, page 48; Cyfarthfa
Castle Museum and Art Gallery, Merthyr Tydfil, page
41; Ironbridge Gorge Museum Trust, pages 20, 25, 29,
30, 45, 47 (right), 49 (bottom), 50 (top); Mick Krupa,
page 10; National Portrait Gallery, page 47 (left); ©
Crown Copyright, Royal Commission on the Ancient and
Historical Monuments of Wales, page 12 (bottom);
Yale Center for British Art, Paul Mellon Collection, pages
14, 22, 42.

The author would like to thank Joanne Smith, of the
Ironbridge Gorge Museums, and Mick Krupa for help in
compiling illustrations for this book.

CONTENTS

THE COMMON METAL

IRON HAS BEEN one of the most important elements in the material world since the Industrial Revolution. As the eighteenth-century ironmaster William Reynolds pointed out, in the iron-bearing rocks below the soil 'lay coiled up a thousand conveniences of mankind'. The iron industry was the driver of much of the economic development of Britain in the eighteenth and nineteenth centuries. To the iron industry we owe the development of large-scale mining in Britain's coalfields, and flourishing secondary industries such as mechanical and civil engineering. British iron was used to construct steam engines, to cast cannon and to lay rails across the world in the nineteenth century. It was the universal metal, the material of the most memorable technical achievements of the day, including the Iron Bridge in Shropshire and the Crystal Palace, as well as an indispensable component of everyday artefacts such as thimbles and mouse traps.

Iron is the commonest of the metallic elements. It unites easily with other elements, especially oxygen (to form rust) and carbon, but also with sulphur and phosphorus, which would provide the greatest technical challenges of the Industrial Revolution period. It has been used since prehistory although it was not the first metal to be exploited. The technology of smelting iron was quite simple, but the technique was difficult, which is why metals such as copper were exploited before iron. However, because of its abundance, once the technical difficulties of smelting it had

The old foundry at Brymbo, near Wrexham, was served by two cupola furnaces for melting pig or scrap iron. These are a rare survival as most iron-industry buildings were cleared when operations ceased, providing sites for redevelopment.

been mastered, it quickly superseded other metals such as bronze and copper.

Iron has been used in three basic forms: wrought iron, cast iron and steel. Wrought iron is pure iron, in commercial if not strictly in chemical terms. When it is white-hot it is malleable and can be hammered (forging) or rolled into a variety of shapes and, depending upon the quality, can be reduced to a thin sheet or narrow piece of wire without breaking. Cast iron has a carbon content of 3–4 per cent and is therefore an alloy of iron. When molten, it can be cast into moulds, but it is not malleable and it is brittle under a hammer, having a granular structure quite different from the fibrous structure of wrought iron. It can be cast into very precise shapes and is strong in compression, which makes it a good building material. Steel has a low carbon content, of between 0.25 and 1 per cent. It is a versatile alloy, capable of being worked like wrought iron, or it can be a more complex alloy containing a number of other elements to give it superior qualities. Chromium and nickel, for example, are added to make stainless steel.

The history of ironworking can be divided into three main phases. The first was the bloomery, a furnace that smelted iron ore to produce malleable iron, which is known as the direct process. It was superseded by the indirect process, whereby iron ore was smelted in a blast furnace to produce molten pig iron, which was then refined in a second process, known as the finery, to produce malleable iron. Alternatively, the molten iron from the blast furnace could be poured into moulds to make cast iron. Until the eighteenth century charcoal was the fuel

The blast furnace at Cefn Cribwr, near Bridgend in South Wales, built in 1771, was initially fuelled with charcoal but later switched to coke.

Moira, Leicestershire, is one of the best-preserved blast furnaces in Britain, but it had a relatively short working life, from 1806 to 1811. Behind the furnace the bridge house is also well preserved, sited next to the Ashby Canal.

used for smelting and refining iron. The introduction of mineral fuel was the first great technological development in ironmaking to occur in Britain. Coke was used in the blast furnace and coal in the various refining processes, bringing a rapid rise in output and a geographical shift of ironmaking to the coalfields. This was the period of the Industrial Revolution, which was followed by the introduction of mass-produced steel, the final phase of the ferrous industries.

The blast furnace at Whitecliff in the Forest of Dean was built against a bank and is the best-preserved part of the former ironworks. The furnace was at work in the first two decades of the nineteenth century.

A—HEARTH. B—HEAP. C—SLAG-VENT. D—IRON MASS. E—WOODEN MALLETS.
F—HAMMER. G—ANVIL.

THE BLOOMERY

IRON WAS FIRST smelted in Britain in prehistory, although the term 'Iron Age' can be misleading if it gives the impression that iron objects are regularly recovered from sites of that period. The Iron Age is the period when iron was first used in Britain, but it remained rare until colonisation by Rome. Iron was used in Greece from about 1200 BC, had reached France and Germany by 700 BC, and soon emerged in Britain. It is a classic case of the slow diffusion of technology from the Near East to the outer reaches of Europe over nearly two millennia.

The bloomery was the small furnace in which iron ore (also known as ironstone or mine) was smelted. Iron ore and charcoal were charged together in the furnace, where waste gases escaped from the top and iron was tapped from the base. Iron ore was first burned to drive off some of its impurities. Charcoal was produced by smouldering wood in heaps, which burnt off impurities, including moisture, to leave a fuel that was almost pure carbon. Charcoal burns hotter than wood, but raising the temperature of the furnace as high as 1,200C was possible only by providing a constant draught of air from hand-operated bellows. The furnace had to be hot enough for the impurities in the ore to melt and form a liquid slag, leaving the iron to coalesce into spongy lumps inside the furnace. After removal of the mass of iron, known as a bloom, at white heat, it was hammered to shape it into a manageable block, and to remove any slag trapped inside the metal. In the Roman period a bloom weighed

This view of a bloomery appeared in *De Re Metallica* by Georgius Agricola, published in Germany in 1556. The furnace is at the top, from where iron was taken to a hammer shown at the bottom.

An experimental furnace, on the scale of prehistoric and Roman period bloomeries, constructed and blown-in in 2010.

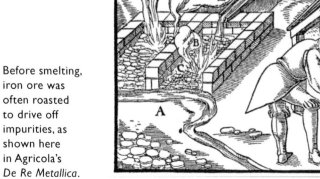

Before smelting, iron ore was often roasted to drive off impurities, as shown here in Agricola's *De Re Metallica*.

about 18 pounds. In practice the bloom needed to be returned to the furnace and hammered several times before the extraneous material had been removed and iron of a tolerable quality was produced (known as bloomsmithing). Then the iron could be sent to a smithy, a separate hearth where it was raised to a white heat and was then hammered to form tools or weapons.

Until the medieval period iron ore was almost invariably dug from the surface. Iron ore is found in many places and iron was worked in areas not now noted for their iron industries. At Bryn y Castell, near Ffestiniog in Snowdonia, iron was smelted for over a century from 100 BC, and then again for another century after AD 150. The ore was probably dug from a peaty mire near the site, since iron ore often forms where iron-rich waters meet organic material. At Brooklands, near Weybridge in Surrey, excavations have revealed evidence of twenty-one furnaces, the result of smelting over a period of five centuries. The best-preserved had a bowl shape dug into the ground, with a clay superstructure, and an internal diameter of 14 inches. On Skye was found a furnace of similar size from the fourth century BC, constructed of stone slabs, with clay used to plug the gaps.

A - Forge   B—Bellows.   C—Tongs.   D—Hammer.   E—Cold stream.

A bloomsmith's forge from Agricola's *De Re Metallica.*

Iron became an important industry in Roman Britain and the chief iron-producing region was the Weald in south-east England. Tools such as knives, as well as hinges, locks, nails, horseshoes and barrel hoops, were forged in its smithies. Larger items such as anchors and beams (used to support copper boilers in bath houses) were forged by welding several

Hematite is one of the richest iron ores found in Britain.

Bryn y Castell is a defended site in Snowdonia, where iron ore was smelted in Iron Age and Roman times. The site still contains waste from the smelting process.

blooms together. Heat-treating of iron embedded in charcoal to produce steel was also introduced in Roman Britain; this was useful for swords because it gave them enhanced strength and hardness.

The British iron industry revived in the medieval period, although for many centuries to come Britain was a net importer

of iron, mainly from the Baltic countries and Spain. Accounts recording the operation of a bloomery at Tudeley, Kent, give an insight into the limitations of ironworking in the fourteenth century. Over a twenty-five-year period from 1329 the bloomery produced between 1.5 and 3 tons of iron a year. Half of the running costs was spent on acquiring fuel, but the main limiting factor in its output was its power supply because it relied on a team of men to work the bellows. The industry solved the problem of power by the introduction of the waterwheel, which could work a pair of bellows and the forge hammers. However, this meant that ironworks now had to be sited next to streams and rivers where there was an adequate supply of water, and this placed it in competition with other energy-hungry industries.

The benefit of using water power is demonstrated by another well-documented works, at Byrkeknott in County Durham, built in 1408. Its waterwheel was capable of blowing the bellows for the bloomery, a second hearth used for reheating the iron known as a string hearth, and for the hammers. The output from its first year of operation was over 24 tons of iron. Even with water power, however, the output of bloomeries remained small when compared to later blast furnaces, but they were still being built after the blast furnace became established in south-east England. A bloomery at Muncaster Head, Cumbria, was built as late as 1636 and remained in production until 1720.

Until the eighteenth century iron was all produced under the hammer. It was an expensive commodity, and decorative ironwork, as seen here surrounding a tomb in the Borbach Chantry chapel in West Dean (Wiltshire), was a sign of high status.

BLAST FURNACE
AND FINERY

IN WHAT WAS known as the indirect process, iron ore was smelted to produce pig iron (an alloy of iron and carbon) in a blast furnace, and was then sent to a forge to be further refined in a finery. The indirect process was probably less an invention than a long process of accident and experimentation. Where the iron ore in the bloomery remained in contact with charcoal at high temperature, iron began to attract carbon. In this form iron has a lower melting point than pure iron, and so pig iron was probably first produced by mistake. A use was found for the molten iron by pouring it into moulds, and cast iron had become part of the metals trade in Europe by the end of the fifteenth century. A more fundamental development was in the refining of this pig iron to produce wrought iron, evidence for which is found as early as the twelfth century in Sweden. This secondary process, the finery, was established in Europe by the fifteenth century, by which time an additional process of heating refined iron ready for hammering into shape was also established and was known as the chafery, counterpart of the string hearth in the bloomery phase.

The first blast furnace in Britain was erected at Newbridge in the Weald in 1496. The Wealden iron industry developed quickly, so that by 1574 it had fifty-two blast furnaces and fifty-eight fineries. Some of the fineries had been adapted from earlier bloomeries. Although the Weald was the largest of the iron-producing regions, the

Paul Sandby's view of a Cumbrian forge is probably of Force Mill, built in 1760. Forges were sited by rivers to exploit the potential for water power.

A small early blast furnace, as depicted by Agricola in *De Re Metallica*.

iron industry developed in other places where there were reserves of iron ore and sufficient woodland for the supply of charcoal. They included the Forest of Dean and neighbouring Monmouthshire, the English Midlands, Yorkshire and Derbyshire, and later on the smaller district of Furness in Cumbria, which had significant deposits of hematite iron ore, which can have an iron content as high as 70 per cent.

Blast furnaces quickly became larger than the bloomeries they superseded. A furnace built at Lydbrook, in the Forest of Dean, in 1635 measured 23 feet high and 23 feet square at the base. At Leighton, in Furness, the blast furnace built in 1713 was served by a 30-foot-diameter waterwheel, driving a pair of bellows each 7½ yards long. These bellows were raised by counterweights and were set to be depressed alternately. This was to ensure that a blast of air remained constant, delivered into the hearth of the furnace by means of a pipe known as a tuyère. At Leighton the furnace was tapped every twelve hours, yielding about a ton of pig iron at each tapping. Seasonal rainfall variations restricted blast-furnace operations to 'campaigns', usually of between six and nine months. In that time the stone lining of the furnace hearth would have been worn away and would need replacing.

The furnace was charged with layers of charcoal, iron ore and limestone. Charcoal and ore were stored in large sheds

With good reserves of raw materials, the Forest of Dean and Wye Valley iron industry expanded from the sixteenth century. Seen here is the seventeenth-century Abbey Tintern blast furnace.

This view of a blast furnace is taken from Denis Diderot's *Encyclopédie*, published in France in 1772. The furnace is charged from the bridge house (K). The large waterwheel powers a pair of bellows inside the blowing house (H). The bellows are depressed by the action of an axle and raised again by counterweights on the outside of the building (I).

Raw materials were stored in large sheds to keep them dry. These eighteenth-century storage sheds are above the blast furnace at Bonawe in Argyll.

above the furnace, and were assembled for charging into the furnace inside a charging house or 'bridge house'. Limestone, containing calcium carbonate, was useful because it combined with the stony constituents of iron ore to form a fluid slag that solidified when it was tapped from the furnace. The iron was run out on the floor of the casting house, usually into a line of ingots known as pigs.

Iron from the blast furnace could also be cast into moulds. The early market for cast iron was for cannon, cannonballs and firebacks. The Weald was the centre of armaments production in the sixteenth and seventeenth centuries – William Levett of Buxted and Stumbletts furnaces in Sussex was described as the 'King's gunstone maker' in 1541. Cannon were cast vertically into pits, but then had to be finished by filing down the inside to achieve the correct bore. This was done by mounting the cannon on a wooden trolley that was advanced upon revolving cutters. It was the beginning of precision engineering.

At the forge pig iron was worked in a charcoal-fuelled finery, its high temperature maintained by the draught from water-powered bellows. The iron was stirred in the hearth to

In the finery, the hearth was fanned by a pair of bellows, on the left of the picture. When the finer removed the ball of iron from the furnace, it was first hammered by hand to remove slag, seen here in the foreground.

ensure that it was exposed to the blast of air, so that oxygen would combine with the carbon in the pig iron and burn it out, leaving the iron in its pure state. Once the iron had formed into a spongy mass, known as a bloom, it was hammered to remove slag and to shape it into a bar, a process known as shingling. The iron was returned to the finery several times before it was refined enough to be sent to the chafery, where

In the forge the bloom was shingled under the hammer. The hammer is raised by trips on a long axletree turned by the waterwheel. Until the advent of steam hammers, all forge hammers were raised in a similar way.

A forge at Dolgellau, Gwynedd, as depicted by Paul Sandby in 1776. Until the industry expanded at the end of the eighteenth century, iron manufacture was a rural industry.

it was brought to a white heat, sufficient for it to be drawn into a bar, in which form it was sold to the market. In practice forges usually had two fineries to provide enough work for a hammer and a chafery.

Since the pig iron from the blast furnace was allowed to cool before it was sent to the finery, the two stages of the indirect process did not need to be done close together. The limiting factor of water supply also sometimes made separation of the two processes a necessity. Blast furnaces were located close to raw materials, but forges were often located closer to markets. Staffordshire and Worcestershire had many more forges than furnaces, with the densest concentration in the Stour valley, where forges could import pig iron via the River Severn and sell bar iron to Birmingham and Black Country manufacturers.

There were three important branches of the iron industry: nail-making, wire-making, and the production of steel for edge tools. One of the most common end uses for iron was in the manufacture of nails. Nails were made from long thin rods formed in a slitting mill; this was like a rolling mill in which a bar, hammered to a flat section, was passed through rolls fashioned with parallel grooves that slit the wide bar into thin rods of square section. Slitting mills proliferated in the Midlands but were found in all of the ironworking districts. Wire manufacture began in the sixteenth century, using technology imported from Germany. As wire was a component of wool cards, it was needed for one of the most important national industries. Iron for wire-making had to be of the highest quality. It required a thin heated rod to be drawn through ever smaller holes, in between which it was annealed, a restorative process in which the iron was heated to 900C and then left to cool slowly, by which means it regained its strength. Steel was manufactured by a process known as cementation, producing an alloy known as blister steel. Iron was packed with charcoal and heated for several days

In the slitting mill, a flat bar was taken out of the furnace, seen here at the back of the picture. Then it was passed through the grooved rolls that slit the bar into rods.

in a sealed chamber, during which carbon from the charcoal infused with the metal, producing a hardened surface, which was important in the manufacture of edge tools. Newcastle upon Tyne and the Derwent valley were at the centre of steel production until overtaken by the Sheffield district in the eighteenth century.

The iron industry was far from being a national industry, because the market for bar iron was regional. Birmingham and the Black Country were served by the works in the Forest of Dean, Staffordshire, Worcestershire and Shropshire. London manufacturers were served by the Weald, via the ports of Newhaven and Rye. In the north, South Yorkshire already specialised in cutlery, supplied by forges in Yorkshire and Derbyshire. So the iron industry had developed a strong regional character by the seventeenth century and it never entirely lost it.

The wireworks at Tintern in the Wye Valley were established by William Humfrey in the sixteenth century. This drawing, made in 1807, shows coils of wire in front of the waterwheel.

THE COKE IRON INDUSTRY

T HERE WAS AN iron industry in Coalbrookdale, near the River Severn in the East Shropshire coalfield, by the mid-sixteenth century, when a bloomery was in operation. It was superseded in 1658 by a blast furnace, the lease for which was acquired in 1708 by Abraham Darby (1678–1717). Darby already had a foundry in Bristol and seems to have taken the lease of the Coalbrookdale furnace to safeguard a supply of pig iron. There was already an established coal trade in Shropshire by this time, supplying domestic fuel to towns on the Severn. It is quite likely that Coalbrookdale furnace men had previously experimented with the use of coke in the blast furnace, perhaps because it had been difficult to find an adequate supply of charcoal. It is certain, however, that when Darby took over its management coke became the sole fuel for smelting. It was not possible to use coal because it contains sulphur, which combines readily with iron and contaminates it. Coke was produced from coal in a similar manner to which charcoal was produced from wood, a process that drove out impurities such

Air, or reverberatory, furnaces were used in the foundry and then the forge in the eighteenth and nineteenth centuries. Heat was deflected from the firebox roof and was drawn over a bowl, in which the iron was worked, by the draught from a tall chimney stack.

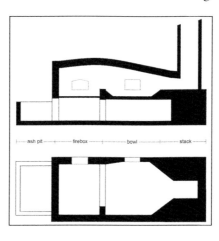

ash pit | firebox | bowl | stack

The fore-hearth of the blast furnace at Coalbrookdale was where the molten iron was tapped. Inscriptions on lintels record the rebuilding of the furnace in 1777 and the original furnace of 1658 (the date was changed in the 1950s by over-enthusiastic campaigners for its preservation).

as sulphur. Coke smelting was not without teething troubles, however, because the right grade of coal for coking was discovered only by trial and error. By 1715 coke smelting had been mastered. We know this because Darby built a second blast furnace at Coalbrookdale, immediately downhill from the old furnace.

The advantage of using mineral fuel was clear. Coal was far more plentiful than wood, which had to be carefully managed, and in the coalfields it was cheap. The number of blast furnaces was no longer limited by the availability of fuel, but of iron ore. Despite these advantages, the iron industry was not experiencing a crisis of fuel supply in the early eighteenth century, Darby never sought patent protection for his innovation, and the use of mineral fuel had little impact for several decades.

Abraham Darby was responsible for another crucial innovation in the iron industry. In the Bristol foundry

Darby's workmen adapted two techniques from the brass industry. One was the use of the reverberatory furnace (also known as an air furnace), in which heat from a firebox was deflected on to a separate bowl and drawn by a tall chimney. Because iron was kept apart from the fuel, it was possible to use coal, the sulphur in which would otherwise contaminate the metal. The air furnace was used to melt pig iron and cast it into moulds. By liberating the production of cast iron from the blast furnace, the air furnace offered much greater flexibility in foundry work. By using more than one air furnace, much larger volumes of molten iron could be cast at a single time than had been possible direct from the blast furnace, allowing for much bigger castings. The other innovation concerned the making of moulds. Until now casting was an expensive business in which the mould was constructed of loam (a mixture of clay, manure, straw

The remains at Bersham ironworks include this eighteenth-century octagonal foundry, a rare surviving building in which multiple air furnaces were built in order to cast items too large for a single furnace.

The 'blowing arch' at Bonawe, where the draught of the bellows was funnelled into a tuyère to blow the furnace.

and sand) and had to be destroyed every time a casting was made. In Darby's foundry wooden patterns were pressed into boxes of sand. This had the advantage that, although the mould would be broken, the pattern could be used many times over. At Coalbrookdale sand-moulding was used initially for manufacturing cast-iron pots, but soon it was also put to other uses, and by 1722 Coalbrookdale was manufacturing steam engines. Cast iron was becoming established as the material of precision engineering.

It was in the East Shropshire coalfield that the coke blast furnace became established. In the 1750s eight new furnaces

The tapping of a coke-fuelled blast furnace at Broseley, Shropshire, in 1788. The crane on the left was used for lifting heavy cast-iron objects when items were cast direct from the furnace.

were built in the area, the first by Abraham Darby II (1711–63), who built new blast furnaces at Ketley and Horsehay in 1754–5. Expansion in Shropshire was made possible by a combination of local expertise and access to a well-established transport infrastructure in the form of the River Severn navigation. On the bank of the river, the Madeley Wood ironworks was established in 1757. Better known as Bedlam, it epitomised the new breed of ironworks. It was established by a group of shareholders, many of whom had been engaged in the coal trade, and was an integrated operation responsible for all stages of production, from getting the raw materials to casting the finished product. The iron and coal trades were henceforth combined.

The coke iron industry also spread further afield in the mid-eighteenth century. In Scotland new coke ironworks were established, such as at Carron (1759) near Falkirk. The biggest expansion occurred in South Wales, where there was abundant iron ore and coal, which were easily obtained as the minerals were found close to the surface at the northern

Bonawe, built in 1753, was one of the last generation of charcoal blast furnaces. The charging house is on the right, and the molten iron would have been run out on to the floor in the foreground.

rim of the coalfield, the area now known as the Heads of the Valleys. Works were established at Hirwaun (1757), and near Merthyr Tydfil at Dowlais (1759), Plymouth (1763) and Cyfarthfa (1765). Nine other works had been founded in South Wales by 1800, including Blaenavon in 1789.

Charcoal iron smelting did not suffer an immediate decline. By 1750 the blast furnaces of the Weald produced less than 10 per cent of the nation's pig iron, but there was considerable new investment in Cumbria in the early eighteenth century, which included new blast furnaces at Backbarrow (1712), Leighton (1713) and Duddon (1736). Cumbrian ironmasters had good reserves of high-quality ore, but, to exploit it fully, they needed to export it to more remote coastal areas where there was plentiful charcoal. So in the 1750s they branched out and established works in the west of Scotland at Bonawe and Craleckan, on the west coast of Wales at Dyfi, and in South Wales at Abercarn.

The growth in ironworks was also a response to an expanding market for iron. This was partly stimulated by the demands of the armed forces in the war-torn years of the mid-eighteenth century, such as the Seven Years War (1756–63),

Dyfi blast furnace, on the west coast of Wales, worked from 1754 to about 1810, fuelled by charcoal. The site has been well preserved and includes a restored waterwheel.

Calcutts ironworks, on the banks of the River Severn in Shropshire, specialised in cannon manufacture during the war-torn years of the late eighteenth century.

during which orders rose rapidly for munitions such as cannon and cannonballs. The precision engineering required in the manufacture of arms was applied to other forms too, notably the casting of steam-engine cylinders. Steam engines were used for mine drainage from the second decade of the eighteenth century, first in the English Midlands, and then further afield. In the 1750s they were also adapted for use at the blast furnace. Output of blast furnaces could be improved if the water supplying the waterwheels could be recycled, so that the supply became more reliable. Pumping engines were installed at several Shropshire ironworks for this purpose. Waterwheels were also now constructed of cast-iron members. A further innovation was the development of the blowing cylinder, replacing large and cumbersome bellows. Ironmasters therefore became key customers of their own innovations.

The most illustrious work to emanate from British foundries of the period was the iron bridge built across the River Severn in 1779; the town that subsequently grew up around it was named 'Ironbridge'. Cast nearby in the Coalbrookdale foundries, the bridge was always intended

In this 1758 view of the Coalbrookdale Upper Works the blast furnace is in the centre and the two taller chimneys are from air furnaces used to re-melt pig iron. In the foreground is an engine cylinder cast at the works.

Opposite: Clydach was one of the generation of ironworks founded in South Wales at the end of the eighteenth century. The blast for the furnace was still powered by a waterwheel in this view of 1811.

to advertise the structural capabilities of cast iron. Although the ironmaster who made it, Abraham Darby III (1750–89), lost heavily in the enterprise, the technology was a triumph and was rapidly improved. Whereas the bridge weighed 378 tons, a new bridge 2 miles upstream at Buildwas, built after devastating floods in 1795, was also made of iron, but weighed only 170 tons. Other innovations followed. At Cyfarthfa in Merthyr Tydfil a bridge was built that carried a water trough, part of a watercourse that supplied one of the works' waterwheels. This concept was developed and scaled up to create cast-iron aqueducts. The bridge at Cyfarthfa was ancestor to a cast-iron aqueduct on the Shrewsbury Canal at Longdon-on-Tern of 1796, and to Thomas Telford's Pontcysyllte Aqueduct in North Wales, completed in 1805.

The market for wrought iron also expanded in the later eighteenth century. Previously, it had been difficult to forge high-quality ironwork from iron smelted with coke. This was partly because coke smelting developed in a part of the Shropshire coalfield where iron ores had a relatively high phosphorus content, an element that makes iron brittle under the hammer. A bigger problem was to find a way of using mineral fuel in the forge. Various methods were tried, but one

This 1798 view of Blaenavon shows the three original blast furnaces and, centre right, the engine house and its tall boiler chimneys. Blaenavon was the first multi-furnace ironworks in South Wales to operate with steam power.

that proved workable and was widely adopted was that known as 'stamping and potting'. Iron was first partially refined in a coal-fuelled finery, just as it had been in the charcoal fineries. It was then beaten into a flat cake and left to cool. When this had happened the iron became brittle and could be broken

Cast in Coalbrookdale in 1779, the iron bridge across the River Severn at the place now known as 'Ironbridge' established the use of cast iron in civil engineering.

up into small pieces (stamping). The granulated iron was then placed in upturned clay pots inside a reverberatory furnace. The pots absorbed heat but prevented contamination of the iron by sulphur in the coal fumes. Charles Wood (1702–74) developed the process at Wednesbury Forge and incorporated it into the Cyfarthfa ironworks, the construction of which he supervised in 1765. The West Bromwich partnership of Wright & Jesson obtained patents for a variant of the technique in 1781 and 1783, and the process was widely adopted across the Midlands.

The increase in wrought-iron production was significant. Stamping and potting helped raise British bar-iron output from 18,800 tons in 1750 to 32,000 tons in 1788, and led to a drift to the coalfields as the iron industry's centre of activity. But the market for bar iron was becoming more complex. Coalfield iron ores tended to produce a relatively hard iron, which was suitable for most purposes, such as the mass production of nails, but was less effective for applications where thin and ductile iron was needed. Wire manufacture, and the manufacture of thin sheets for tinplating, continued to rely on the charcoal forge.

PUDDLING AND ROLLING

THE TECHNIQUE THAT allowed mass production of wrought iron in the nineteenth century was known as puddling. Henry Cort (1740–1800), a Navy agent in Portsmouth, conceived the idea that British bar iron could compete with high-quality Baltic iron for lucrative contracts with the Royal Navy. Cort acquired a small forge at Fontley in Hampshire, where the workmen developed a radical new way of refining pig iron, for which Cort was awarded patents in 1783 and 1784. In Cort's method, only later known as puddling, the iron was stirred in a reverberatory furnace to ensure that all of the iron was exposed to oxygen, and that the carbon was all burnt out. The other innovation was that the iron was then passed through rollers, which proved very effective in removing the slag from the puddled ball of iron. Introduction of rollers would prove crucial. Aided by the development of the steam engine, rolling mills would become standard plant in ironworks, where they had previously been confined to specialist work such as rolling plates. Rolling mills produced more consistent sizes and shapes than was possible under the hammer, and more quickly too.

Cort's efforts to interest the iron trade in his new process were largely unsuccessful. One ironmaster who did agree to adopt the process, and to pay Cort a royalty for its use, was Richard Crawshay (1739–1810). Crawshay was a successful London iron merchant, but was new to iron manufacture. He had taken over the small ironworks at Cyfarthfa, for which he had high ambitions. Cort's process suffered from teething

In a calcining kiln ironstone was roasted to remove some of its impurities. The kiln was charged with coke and ironstone at the top and drawn at the base. These rare surviving kilns are at Tondu in South Wales.

problems, however, and before they were sorted out Cort had been declared bankrupt, and thereby forfeited his patents. By 1793 puddling was a success at Cyfarthfa, and at a neighbouring ironworks, Penydarren, and it soon came to be known as the 'Welsh method'. The iron industry in South Wales was growing fast and it soon became the leading iron-producing region in Britain – in 1805, 30 per cent of British pig iron was smelted in the blast furnaces of the South Wales coalfield.

Puddling was adopted across all of Britain's ironmaking regions. South Wales was dominated by integrated ironworks, where iron was smelted in blast furnaces and then refined in adjacent forges. In the Midlands, which already had a well-established market for bar iron, there were also integrated ironworks, but the older organisational system still thrived. Independent forges continued to purchase pig iron on the open market. Some of these forges continued to refine pig iron in charcoal fineries, producing high-quality iron such as wire, for which there was a growing market. In practice, many of the coalfield ironmasters purchased older charcoal forges because it allowed them to offer a greater range of iron products. By the 1830s some of these works, including

A view of Penydarren Ironworks, Merthyr Tydfil, in 1811. The forge on the right side of the picture has the chimneys of puddling furnaces, which were customarily built in pairs.

Horsehay and Old Park in Shropshire, incorporated a charcoal iron department to supplement their puddling forges.

The other major innovation of the early nineteenth century was the introduction of hot blast, which was patented in 1828 by James Neilson of Glasgow. Hitherto it had been received wisdom that blowing cold air into the blast furnace made the best quality of pig iron, on the basis that the furnaces were more productive in the colder winter months than in the summer. However, it was found that hot blast significantly reduced the quantity of fuel needed for smelting. Hot blast had a significant impact on the geography and output of the British iron industry because new ironworks were established in areas where there were exploitable reserves of iron ore, but where development had been inhibited by relatively modest coal reserves. In Scotland in 1830 twenty-four blast furnaces smelted 37,500 tons of pig iron. By 1839 there were fifty-four Scottish blast furnaces, mostly using hot blast, with an output of 195,000 tons. The rapid development of smelting in

Blast furnaces at Etruria, Staffordshire, in 1872. Pipes leading from the top of the furnace show how waste heat was recycled to heat the air blown into the furnaces at the level of the hearth.

This beam blowing engine at Ebbw Vale Ironworks was manufactured by the Perran Foundry in Cornwall. The power cylinder is on the right and the blowing cylinders on the left of the pivot.

The blast furnace was charged at the top. Until the nineteenth century raw materials were tipped in from baskets, but these were gradually superseded by barrows as the size of furnaces increased.

Cleveland occurred after 1850 for the same reason. Cleveland had 104 blast furnaces by 1872 and had become Britain's leading producer of pig iron.

We know from contemporary accounts how an ironworks operated in the mid-nineteenth century. As they were charged at the top and tapped at the base, there were two working levels at a blast furnace. At the upper level was the coke yard. Coal was converted to coke either by burning it in open heaps, or by heating it in brick ovens. 'Mine burning' was conducted in open heaps, or in purpose-built 'calcining kilns'. In the eighteenth and nineteenth centuries it had been customary to assemble the raw materials in a charging house at the head of the furnace – a throwback to the days when it was important to keep charcoal dry – but these gradually fell from favour.

Once they had been lit, the

furnaces had to remain in blast, because if a furnace cooled too much the lining would crack. Men worked in twelve-hour shifts six-and-a-half days a week, with a half-day off to allow maintenance work. Gangs of fillers charged the furnaces, but the responsibility for ensuring that it was charged in the correct proportions was the job of the furnace keeper. The key event was the furnace tapping every twelve hours, when the blowing engine was stopped. Because it was lighter, the slag floated on top of the molten iron at the hearth of the furnace, and was tapped first. The furnace keeper could deduce from the consistency of the slag whether or not the furnace was working efficiently. If the slag was too viscous, then more limestone was needed; if it was too fluid, then more ironstone was needed. The molten iron was run out into sand beds to form ingots, known as pigs.

Before it was sent to the puddling furnace, pig iron was melted again in a 'running-out fire' in order to remove its silica content, although use of these declined from the 1830s when the puddling process was improved by a technique known as pig boiling. The running-out fire was a simple

The former engine house at Llynfi Ironworks in Maesteg was built in 1839. It has the simple but sturdy elegance typical of nineteenth-century engine houses.

furnace with a tall stack, in which the pig iron was heated. When it began to melt, a blast of air was introduced and after about two hours the molten pig iron was ready to run out into moulds. The resulting 'finers' metal' or 'plate' was then broken up ready for the puddling furnace.

Puddling was carried out in a reverberatory furnace (see page 25), which never changed radically from its original form. The iron in the furnace was worked through a small opening in the furnace door. The puddler stirred the iron with a long bar, or rabble, to ensure that all of it was exposed to the air and that none of it adhered to the furnace bottom. The heat would take up to one and a half hours, with a charge of about 4 hundredweight. Burning out of the carbon was signalled by a blue flame, whereupon the iron was stirred more vigorously and slowly formed into pasty masses, known

Ynysfach engine house in Merthyr Tydfil was built in 1836 to provide the blast for two furnaces. It has the tall, narrow dimensions that distinguish buildings erected to house beam engines.

The puddler removes a ball of iron from the puddling furnace. The furnace door is opened by means of a chain.

in the trade as 'coming to nature'. The iron was removed in five or six balls, and then a tap hole was unplugged to allow the slag to run out.

The iron was worked under a hammer to shape it into a rough block, sufficient for it to be passed through a rolling mill, where the remaining slag was squeezed out. Puddled bars were then cut up and stacked, or 'piled', in another reverberatory furnace, known as the 'balling furnace', where they were brought to a white heat and then passed through a rolling mill. Depending on the grade required, the process of balling and rolling could be repeated more than once. In general, the more the iron was worked, the better its quality – which is why scrap iron was not regarded as a waste product but was capable of producing premium-grade iron.

Work on the rolling mill was labour-intensive and required dexterity and physical stamina to run the iron through the rollers correctly. When the rollerman had passed the bar through the rollers, the catcher passed it back over the top

The rolling mills and balling furnaces at Cyfarthfa Ironworks, painted in 1825. In the distance on the right of the picture is Cyfarthfa Castle, newly built by the ironmaster, William Crawshay II.

of the rollers – a 'dead pass' – for the rollerman to pass it through the next smallest groove of the rollers. This method was gradually superseded in the nineteenth century, first by 'three-high mills', in which the bar was passed through the grooved rolls in both directions, and finally by 'strip mills', in which bars were guided through consecutive pairs of rolls, reducing the labour, and skill, required.

By the mid-nineteenth century most rolling mills were steam-powered, but not exclusively so. At Cyfarthfa, water power was used in the mills for as long as the rolling mills were in work. The advantage of steam-powered mills was that they could be erected anywhere and they could provide sufficient power for heavy work. The railway boom that began in the 1830s encouraged use of more powerful engines than had hitherto been required for rolling bar. In 1857–9 the new Goat Mill was built at Dowlais for rolling rails; according to the engineer William Menelaus, it had three times the power of any existing

Traditional forge hammers were raised by trips on an axle and fell by gravity. They could be worked by waterwheel or steam engine.

engine, even though it was still a traditional beam engine. Steam power was also applied to forge hammers. Where forges were rolling bar iron in small sections, a traditional hammer raised by the trips on an axle and then falling by gravity was fine, and was preferred by many shinglers. For heavier work, or forges where a hammer was required to service up to thirty puddling furnaces, then steam hammers were preferred and were widely used in the iron industry.

In the nineteenth century the market for iron diversified and specialised, so that almost every ironworks served only a specific sector of the market. In practice the iron industry was a series of regional industries, often serving quite different customers. In the railway boom years of the 1830s and 1840s ironworks were busy forging rails, but it was a high-volume business that demanded investment in powerful engines and rolling mills specific to the job. Many of the South Wales ironworks specialised in rail production. Dowlais produced over 2,000 tons of rails by 1836 and went on to supply national and international markets. In 1839 the smaller Pentwyn & Golynos Iron Company of Monmouthshire won orders for 1,500 tons of rails for the Taff Vale Railway and 14,000 tons for export to Russia, orders that would have occupied the works for over a year.

In the Midlands there was more concentration on the traditional bar-iron market, producing iron that was sold to manufacturers to make into general or specialised wares.

A steam hammer re-erected at Blaenavon Ironworks.

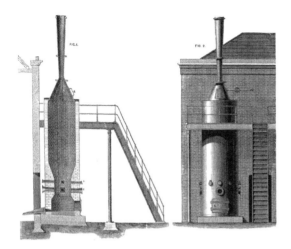

For this reason the Midland forges were generally relatively small in scale. Charcoal iron was still in demand for manufacture of wire, not just for fencing, but for telegraph wires and for small items such as springs and bottling wire – the wire used to tie down the corks of soda-water bottles, 100,000 miles of which was manufactured in 1866. The railway boom also brought demands for new grades of iron for components used in locomotive manufacture. West Bromwich became a centre for manufacturing iron springs and couplings from the 1830s, a move to sophisticated engineering products for an area that had previously specialised in the more primitive work of nail-making.

Cupola furnaces are tall brick furnaces clad in iron or steel sheets. They are charged at the top and tapped at the base. They have been in use since the end of the eighteenth century for foundry work.

Charcoal-fined iron was also used in the manufacture of tinplate, which was a specialised branch of the iron industry. Forges in the English Midlands had pioneered its use but by the mid-nineteenth century tinplate manufacture was dominated by works in South Wales, but not overlapping very much with the rail-making forges. Tinplate manufacture was concentrated in the west of the coalfield, the Neath and Swansea valleys, extending into Carmarthenshire further west. To manufacture tinplate, iron was rolled to a thin section, annealed and cleaned, and then dipped in molten tin.

Engineers such as Isambard Kingdom Brunel, John Napier and John Laird began building iron-hulled ships in the 1830s and 1840s. Scotland's forges specialised in the manufacture of ship plates, boiler plates, and other marine fittings such as crank and propeller shafts. Ironworks in Cleveland also specialised in marine engineering and moved into the rail market previously dominated by South Wales. Some Yorkshire firms, including the Park Gate Works near Rotherham, specialised in armour plating, and by the 1850s were rolling massive plates 4 inches thick for the early ironclad warships.

Because of the nature of the iron ores in the East Shropshire coalfield, some of the ironworks there specialised in smelting pig iron for sale to foundries. Cast iron found many new applications in the nineteenth century, although it was still used regularly in bridge construction. Reverberatory furnaces were small, however, and a new type of furnace was needed in order to increase the capacity of the foundry. The cupola furnace was patented by William Wilkinson in 1793. It is still widely used. The market for cast iron expanded in the nineteenth century and later, as foundries produced cast iron for all types of engine. Cast iron even found favour in architecture and ornament. Cast-iron gates and fountains in municipal parks, cast-iron pillar boxes and telephone kiosks all became familiar sights in British towns and cities. Many companies specialised in their manufacture, such as Coalbrookdale, Macfarlane & Company of Glasgow and Andrew Handyside of Derby.

Tapping of a cupola furnace in 1994 at a foundry in Walsall, where small items of harness-ware were cast. The molten iron was poured from the ladles into casting boxes.

MASTERS AND MEN

WHEN THE FIRST blast furnaces and fineries were established in the Weald workmen from Normandy were hired to work at them. They were the only people who had the necessary skills. Moreover, the iron industry was in many respects a closed community, a male-dominated world in which skills were handed down from father to son. Among the Wealden ironworkers' families were the Lavenders, whose descendants could be found working in the Midlands iron industry in the eighteenth and nineteenth centuries, and there were many others like them.

The ironmaster was the man responsible for running an ironworking business, although in practice he was normally only joint owner of it. The ironmaster seldom had any practical ironmaking skills of his own. The position of ironmaster tended to become dynastic – the Homfray family remained ironmasters across Staffordshire, Shropshire and South Wales from the seventeenth to the nineteenth centuries – and family concerns predominated until the 1862 Companies Act. Some ironmasters began as workmen, such as John Guest of Broseley, Shropshire, whose descendant Sir Josiah John Guest would become one of the richest industrialists in Britain in the nineteenth century. Some ironmasters were engaged in related businesses, such as Richard Crawshay, who was an iron merchant, or Thomas Botfield, who was a land agent, both of whom gained a privileged insight into the potential for iron manufacture

Opposite left: Sir Josiah John Guest (1785–1852) became one of the wealthiest of nineteenth-century ironmasters, and his ironworks at Dowlais grew to be the largest in Britain. He was grandson of John Guest, an ironfounder from Broseley.

and made the most of it. Where an ironworking concern was spread across more than one site, for example furnace and forge, managers were appointed under the ironmaster, many of whom, like John Wheeler, who worked for the Foley family which dominated the Forest of Dean iron industry in the seventeenth century, later became ironmasters themselves. The great irony, however, is that the ironmasters and entrepreneurs who introduced the most radical changes in the industry – Abraham Darby, Henry Cort, Richard Crawshay and Sir Henry Bessemer – were all newcomers.

Ironmasters made the capital investment and provided all the plant and tools for the job; the commodity that workmen traded was their skill. The iron industry relied upon, in Richard Crawshay's phrase, 'active and powerful men'. Puddlers and rollermen learned their trade from boyhood, were active in it during their peak years, and in older age took on lighter duties. Few puddlers retained the strength and stamina required for the job once they had reached fifty. Skill

Above right: William Reynolds (1758–1803) was ironmaster of Ketley and Bedlam ironworks in Shropshire and pioneered the use of cast iron for canal aqueducts.

in the iron trade was passed on by, and effectively controlled by workmen, often to the chagrin of ironmasters. The initial attraction for ironmasters of puddling and similar techniques was the vain hope that they might obviate the need to hire skilled men.

Traditionally all workmen were paid piece rates and were allowed to employ whomever they chose to help them. In the boom years skilled workmen were very well paid, but their prosperity was at the mercy of market forces. Ironmasters did not stockpile their wares, so the workmen suffered, and many lost their jobs in times when output decreased, such as the peace that followed the Battle of Waterloo in 1815.

Workmen enjoyed customary privileges, such as an allowance of beer and a place to live. Large ironworks employed the services of a surgeon to attend the injured – usually victims of burning or of machine-related accidents. Not all of these medical men were diligent in their duties, however. As ironworks were usually set up outside established

Cyfarthfa Castle was built in 1825 for William Crawshay II. The imposing mansion overlooked the ironworks, and its lake doubled as a reservoir for the works.

Butetown is a planned village built for workmen at Union Ironworks near Rhymney in 1802–4.

In this view of Coalbrookdale published in 1758, houses are ranged on the hill above the works reservoir. The houses include two built for ironmasters, and Tea Kettle Row, a terrace of workmen's houses begun in 1735.

towns, ironmasters had often needed to offer houses as an inducement to skilled workmen, and in the rapid expansion of the industry from the end of the eighteenth century houses built by ironmasters and by speculative builders created towns such as Merthyr Tydfil, Tredegar and Rhymney.

A trade token issued by John Wilkinson who owned works at Bersham, Brymbo (both at Wrexham), Bradley (Staffordshire), Snedshill and Willey (both Shropshire).

Not all workmen lived in a terraced house. In this engraving of Blaenavon published in 1801, makeshift homes can be seen beneath the arches of a viaduct.

Workmen's houses varied enormously in their standard of accommodation, and until the nineteenth century were often built within the works itself. Some ironmasters paid their workforce in tokens that could be redeemed only in company shops and beer houses – the so-called truck system that was widely resented by workmen and easy for employers to abuse by charging inflated prices. Ironmasters founded educational institutes for workmen and provided schools for their children. But the institutions that are most closely associated with the Industrial Revolution were beyond the influence of ironmasters: the chapels that fulfilled the spiritual, social, cultural, educational and independent aspirations of working men.

STEEL

IRONMASTERS WERE LOOKING for a cheaper and quicker way to mass-produce iron than the puddling furnace. The puddling furnace in 1860 was no larger than it had been in 1790. In that time blast furnaces had grown ever larger, with ever more complex ways of improving their efficiency, for example by means of recycling waste heat. If ironmasters wanted to increase their pig-iron output they could build larger blast furnaces, but to increase bar-iron production they had to build more puddling furnaces and employ more puddlers. The strip mill successfully mechanised the rolling of iron bars, but attempts at mechanisation of the puddling process all failed because the process still relied upon the judgement of the puddler.

An early Bessemer converter. Molten iron was poured into the top (as seen on the left) and a stream of cold air was blown through air holes in the base, before it was rotated to pour the molten metal into moulds.

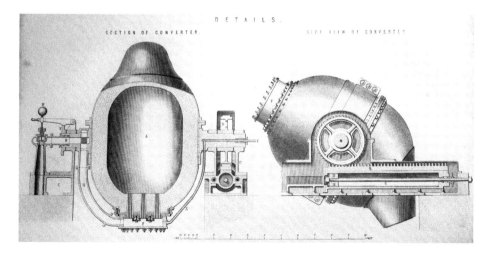

ELEVATION.

Beam blowing engines were still favoured in the later nineteenth century because of their reliability. This example from Carnforth Hematite Works in Lancashire has two power cylinders and consequently two beams.

The answer to the ironmasters' problems came in the manufacture of steel. Hitherto steel had been difficult and expensive to produce. In the eighteenth century Benjamin Huntsman found a new way of manufacturing steel, but it required the iron to be melted at a very high temperature, which was difficult to do on a large scale. The solution was to melt the iron in small crucibles – hence 'crucible steel' – in a branch of the iron industry centred on Sheffield.

Two mid-nineteenth-century innovations revolutionised the iron industry: the Bessemer converter, and the open-hearth furnace. In Henry Bessemer's process molten pig iron was run into a large vessel, through which a stream of air was passed. The cold air burned out the carbon in the metal, which was then cast into an ingot. It produced mild steel, with a carbon content of up to 0.3%, capable of being shaped in a rolling mill, just like wrought iron. The process attracted immediate interest, but success was patchy at first. Sheffield was at the forefront of developing Bessemer steelmaking – it had twenty-seven converters in 1871 and thirty-six by 1878. But the process did not work where phosphoric ores were smelted, because the phosphorus content was not removed. This problem was solved by Sidney Gilchrist Thomas in Blaenavon. He lined the converter with brick made of dolomitic limestone and a fireproof tar mixture. This was known as the 'basic' Bessemer process, distinguished from the original 'acid' process. Unfortunately it did little to halt the decline of iron mining in South Wales.

The open-hearth furnace was patented by Frederick Siemens in 1856, and was improved in 1861 when an associated gas producer obviated the need to use solid fuel. Gas producers produced combustible gas, similar to a coal-gas retort, using inferior and cheap grades of coal. The furnaces worked on a regenerative principle: when the furnace had

Barrow Hematite Steelworks, viewed in the 1860s, was one of the most modern plants in Britain at the beginning of the steel era.

BLAST FURNACES AT BARROW-IN-FURNESS.

This view of the fore-hearth of number 5 furnace at Cyfarthfa shows that it was re-lined in 1879 ready for blowing-in. In the event the furnace was never used again because new furnaces were erected for manufacturing Bessemer steel.

heated up, exhaust gases were passed through a brickwork chamber, which effectively heated it up too. Then the draught was reversed and the combustible gases were passed through the heated chamber, where they ignited, while the exhaust was passed through the opposite 'cold' chamber. By continually reversing the draught, the furnace could be maintained at a temperature of between 1,500C and 1,800C. Open-hearth furnaces were a cheap way of making malleable steel using a bath of molten pig iron in which scrap, notably old iron rails, was melted.

Once steel was mass-produced commercially it rapidly proved to be superior to wrought iron for most applications. Sheffield steelmakers were quick to adopt steel for rails as steel rails were more durable than iron rails. By 1873,

250,000 tons of steel rails were rolled in Sheffield, and the biggest firm, the Cyclops Works, won orders from railways in Britain, the United States, Canada and South America. In marine engineering steel ship plates prevailed slightly later, but by 1888, 91 per cent of ship plates manufactured in Britain were steel. In the 1880s steel also replaced wrought iron in tinplate manufacture.

In the older markets for wrought iron, steel did not supersede iron until after 1900. For some uses resistance to corrosion was important, so for chains, cable and ships' nuts and bolts wrought iron was still preferred up to 1914. The British wrought-iron industry nevertheless declined in the final quarter of the nineteenth century. British ironworks were undercut by cheaper iron from Belgium and Germany. Many of these declining works in the Midlands and South Wales were long established. Their plant was old, but investment in modernisation was not inevitable. Many of the best ironstone reserves were nearly worked out, necessitating imports of iron ore, which favoured coastal works. When the ninety-nine-year lease of the Cyfarthfa ironworks was renewed in 1864 William Crawshay II did so only reluctantly. In practice Cyfarthfa, and many other works founded at the end of the eighteenth century, would soon give up the iron trade. Many of the Welsh ironmasters possessed extensive mineral rights, and mining and selling coal was more profitable for them than investing in iron production. This was equally true in Shropshire, where fireclay was also mined to manufacture bricks, a simpler process than iron production, and more profitable.

Some of the older ironworks did invest in steelmaking, notably Dowlais, although in 1891 the company invested in a new steelworks at East Moors in Cardiff, which signalled the demise of mass steel production at the head of the Taff valley above Merthyr Tydfil. New areas now prospered. The discovery of iron ore in Cumbria was behind the success of

Clydach
Ironworks,
Monmouthshire,
was one of the
generation of
South Wales
ironworks
that declined
and ceased
production in
the 1870s.

the Barrow Hematite Steel Company, established in 1850 to smelt pig iron, using coal imported from County Durham.

Bessemer and Siemens introduced mechanical processes that no longer relied upon the judgement of workmen. Steel had become a technical rather than a craft industry. Technology could be disseminated quickly, however, and in the twentieth century Britain ceased to enjoy technological and economic supremacy in the ferrous industries. The global steel industry drew much from Britain, however. The basic Bessemer process devised by Sidney Gilchrist Thomas had some benefit to British producers, but arguably its real beneficiaries were Andrew Carnegie in Pittsburgh and Alfred Krupp in Essen.

PLACES TO VISIT

THE SURVIVING HERITAGE of the iron industry in many ways gives a skewed impression of the industry as a whole. It is confined largely to works of the eighteenth and nineteenth centuries, and to blast furnaces rather than forges. Blast furnaces encased in thick masonry shells have often remained standing by virtue of their indestructibility, while the ancillary buildings around them were either demolished when a works closed or have slowly fallen down. The survival of machinery *in situ* is also extremely rare. The list below is confined to sites that are either open to the public or can be viewed from public places.

Abbey Tintern Furnace, Tintern, Monmouthshire. Remains of an eighteenth-century blast furnace and associated buildings, as well as a series of ponds that served forges and wire works.

Anne of Cleves House, 52 Southover High Street, Lewes, East Sussex BN7 1JA. Telephone: 01273 474610. Website: www.sussexpast.co.uk
Although there are no surviving blast furnaces in the Weald, this museum has collections encompassing the products of Wealden ironworks, including cannon and firebacks.

Bersham Heritage Centre and Ironworks, Bersham Road, Wrexham LL14 4HT. Telephone: 01978 318970. Website: www.wrexham.gov.uk
An important site from the coke era.

Blaenavon Ironworks, North Street, Blaenavon, Torfaen NP4 9RN. Telephone: 01495 792615. Website: www.cadw.gov.wales
The best-preserved ironworks in South Wales.

Bonawe Historic Iron Furnace, Taynuilt, Argyll PA35 1JQ. Telephone: 0131 668 8600.

Website: www.historic-scotland.gov.uk
An outstanding example of a charcoal-fuelled
blast furnace.

Brymbo Steelworks, Brymbo, Wrexham. An important
coke-era site.

Bryn y Castell, near Ffestiniog, Gwynedd. This prehistoric
ironworking site in North Wales has been restored.

Cefn Cribwr Ironworks, Bedford Park, Cefn Cribwr,
Bridgend.

Clearwell Caves, The Rocks, Coleford, Gloucestershire
GL16 8JR. Telephone: 01594 832535.
Website: www.clearwellcaves.com
Tells the story of ironstone mining in the Forest of Dean.
Clydach Ironworks, Gilwern, Brynmawr.
Website: www.breconbeacons.org/clydach-ironworks
Craleckan Ironworks, Furnace, near Inveraray, Argyll.
Historic blast furnace.
Cyfarthfa Castle Museum and Art Gallery, Brecon Road,
Merthyr Tydfil CF47 8RE. Telephone: 01685 727371.
Website: www.visitmerthyr.co.uk
Built as the home of the ironmaster William Crawshay
II in 1825, the museum explains the history of Merthyr's
iron industry. The nearby remains of the Cyfarthfa
Iron and Steel Works have the most impressive bank of
surviving furnaces in Britain.
Cynon Valley Museum and Gallery, Depot Road, Gadlys,
Aberdare CF44 8DL. Telephone: 01685 886729.
Website: www.cvmg.co.uk
Incorporates part of the Gadlys ironworks.
Dean Heritage Centre, Soudley, Cinderford,
Gloucestershire GL14 2UB. Telephone: 01594 822170.
Website: www.deanheritagecentre.com
Tells the history of ironworking in the Forest of Dean.
Duddon Iron Furnace, Broughton-in-Furness, Cumbria. The
best-preserved charcoal blast furnace complex in England.
Dyfi Furnace, Furnace, near Machynlleth, Powys (Cadw).
Telephone: 01443 336000. Website: www.cadw.gov.wales
The best-preserved charcoal blast furnace complex in Wales.
Ironbridge Gorge Museums, Telford,
Shropshire. Telephone: 01952 433424.
Website: www.ironbridge.org.uk
Cares for the biggest concentration of surviving
ironworking sites in the Midlands. They include
the following.

Opposite: The
first blast furnace
at Brymbo, near
Wrexham in
North Wales,
was built in
1796 by John
Wilkinson and,
together with
its adjoining
casting house,
is undergoing
restoration.

Bedlam Furnace, Waterloo Road, Ironbridge. To the east of
Ironbridge, set back from the bank of the River Severn,
are the furnaces of the Madeley Wood Company.

Blists Hill Victorian Town, Legges Way, Madeley, Telford,
Shropshire TF7 5DU. The open-air museum incorporates
another furnace complex of the Madeley Wood Company.
It also has reconstructed puddling furnaces and a steam
hammer, and a large blowing engine of 1851 brought
from nearby Priorslee.

Coalbrookdale Museum of Iron, Dale Road, Ironbridge,
Telford, Shropshire TF8 7DQ. The site of the former
Upper Works of the Coalbrookdale Company includes
a blast furnace built in 1777 and the museum in a
warehouse of 1838.

Llynfi Ironworks, Maesteg, Bridgend.

Two blast
furnaces survive
at Bedlam, by
the River Severn
near Ironbridge,
both built in
the nineteenth
century. One
is behind the
masonry casing,
and the other,
on the right, was
a freestanding
brick structure.

Moira Furnace, Furnace Lane, Moira, Swadlincote, Derby DE12 6AT. Telephone: 01283 224667. Website: www.moirafurnace.org
Well-preserved furnace.

Neath Abbey Ironworks, Neath, Port Talbot, Wales.

Parc Tondu Victorian Ironworks, Tondu, Bridgend CF32 9TF. Telephone: 01656 722315. Website: www.visitwales.com

Rockley Furnace and Engine House, Rockley Lane, Worsbrough, Barnsley (South Yorkshire Industrial History Society). A notable example of a charcoal-fuelled blast furnace.

Rural Life Centre, The Reeds Road, Tilford, Farnham, Surrey GU10 2DL. Telephone: 01252 795571.
Website: www.rural-life.org.uk
Has a half-scale reconstruction of a Wealden blast furnace.

Whitecliff Furnace, Newland Street, Coleford, Gloucestershire GL16 8ND. Website: www.fodbpt.org

The best-preserved blast furnaces are those that were encased in masonry, as seen here at Llynfi, near Maesteg, where the base of a furnace built in 1837 has survived.

Below the furnace at Abbey Tintern is a reservoir and leat that supplied the waterwheel at one of the wireworks further down the valley.

Wortley Top Forge, Forge Lane, Wortley, Sheffield S35 7DN. Telephone: 0114 281 7991. Website: www.topforge.co.uk Retains cranes and tilt hammers in addition to restored waterwheels.

Ynysfach Engine House, Ynysfach, Merthyr Tydfil CF48 1AR. Telephone: 01685 725039. Website: www.visitmerthyr.co.uk

FURTHER READING

Bowden, Mark. *Furness Iron*. English Heritage, 2000.

Dawson, Frank. *John Wilkinson: King of the Ironmasters*. The History Press, 2012.

Evans, Chris. *The Labyrinth of Flames: Work and Social Conflict in Early Industrial Merthyr Tydfil*. University of Wales Press, 1993.

Gale, W. K. V. *The Iron and Steel Industry: A Dictionary of Terms*. David & Charles, 1971.

Hay, Geoffrey, and Stell, Geoffrey. *Monuments of Industry: An Illustrated Historical Record*. HMSO, 1986.

Hayman, Richard. *Ironmaking: The History and Archaeology of the Iron Industry*. Tempus, 2005.

Hodgkinson, Jeremy. *The Wealden Iron Industry*. The History Press, 2008.

Ince, Laurence. *The Knight Family and the British Iron Industry*. Ferric Publications, 1991.

Ince, Laurence. *The South Wales Iron Industry 1750–1885*. Merton Priory Press, 1993.

Meredith, Laurence. *The Iron Industry of the Forest of Dean*. Tempus, 2006.

Riden, Philip. *A Gazetteer of Charcoal-Fired Blast Furnaces in Great Britain*. Merton Priory Press, 1993.

Shill, Ray. *South Staffordshire Ironmasters*. The History Press, 2008.

Sim, David. *The Roman Iron Industry in Britain*. The History Press, 2012.

Strange, Keith. *Merthyr Tydfil, Iron Metropolis: Life in a Welsh Industrial Town*. Tempus, 2005.

INDEX

Printed and bound by CPI Group (UK) Ltd, Croydon, CR0 4YY

11/10/2024

01043563-0016